DINOSAUR DISCOVERY

LOST CREATURES OF THE CRETACEOUS

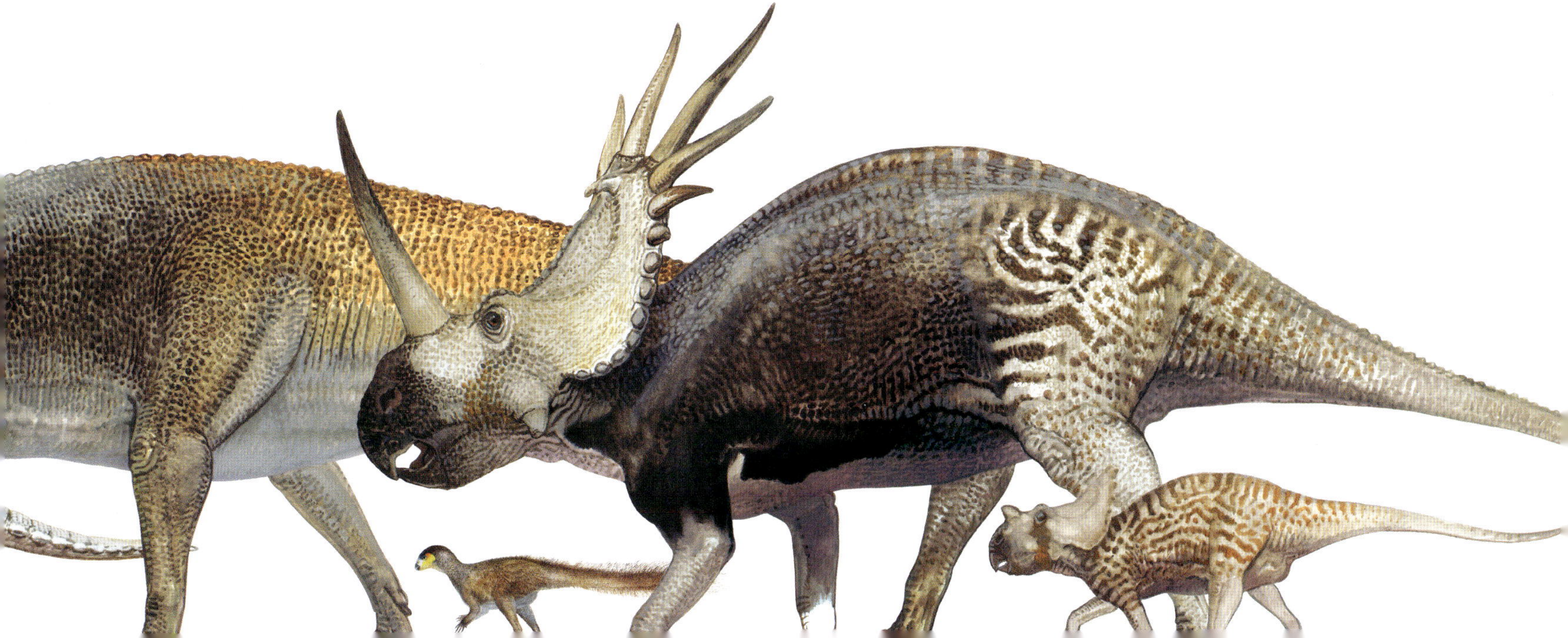

Published 2014 by the Western Australian Museum
49 Kew Street, Welshpool, Western Australia 6106
(Postal: Locked Bag 49, Welshpool DC, WA 6986)
www.museum.wa.gov.au

Designed by Tim Cumming.
Printed by Everbest Printing Company, China.

ISBN 978-1-925040-12-8

BRING THE DINOSAURS TO LIFE

Download the free dinosaur discovery app for your apple® or android® device from the app store or google play.

LOOK FOR THESE MARKERS

Scan these symbols using the app and watch your very own dinosaurs come alive.

CONTENTS

FOREWORD

145 million years ago, the Earth was a very different place to the one we inhabit today. If we could travel back in deep time, we would barely recognise Cretaceous Australia with its volcanoes, shallow inland seas and great rift valleys stretching across vast distances.

This souvenir book has been developed by the Western Australian Museum, in conjunction with its hugely popular *Dinosaur Discovery: Lost Creatures of the Cretaceous* exhibition. Authoritative, interactive and entertaining, this guide reveals the latest details of how scientists believe these creatures looked, behaved, hunted, defended themselves, and cared for their young — there is even a section on dinosaur droppings.

This guide describes and illustrates the dinosaurs of the Cretaceous era, the amazing environment in which they lived, and how evidence of them became preserved as fossils. It also explores the evolution of the Earth during that time, the cataclysmic asteroid strike that probably brought the reign of the dinosaurs to an end, and the subsequent rise of mammals as dominant land animals.

The exhibition, *Dinosaur Discovery: Lost Creatures of the Cretaceous*, was created by the Western Australian Museum with Goldie Marketing and is enjoyed by audiences of all ages. I hope they enjoy this guide too, as they encounter some of the most incredible creatures ever to have walked the Earth.

Alec Coles OBE
Chief Executive Officer
Western Australian Museum

INTRODUCTION

The 'Age of the Dinosaurs' began 230 million years ago in the Triassic Period and by Jurassic times dinosaurian household names like *Allosaurus* and *Brachiosaurus* had evolved into the giant creatures we have become so familiar with. Dinosaurs reached their peak in diversity during the following Cretaceous Period which began 145 million years ago.

The Cretaceous climate was greatly different to that of our's today. It was a greenhouse world where the polar regions were largely or entirely ice free and sea levels were generally much higher, resulting in the fragmentation of continental landmasses by vast inland seas.

Spinosaurus and other giant dinosaurian carnivores of the Cretaceous Period were the ultimate guardians of this ancient world. Along with smaller but swift theropods like *Velociraptor* and *Australovenator*, they effectively put a lid on the evolution of large mammals. Our small furry ancestors were confined to a life in the shadows until the non-avian dinosaurs were gone. No mammal was larger than a Tasmanian Devil during the Cretaceous Period and most were much smaller.

The end of the Cretaceous, 66 million years ago, changed everything. A huge asteroid or comet slammed into our planet at incredible speed and caused total devastation. All dinosaurs died out except for one type of feathered, avian maniraptor (more commonly known as birds), giving way to the eventual rise of mammals as the dominant land vertebrates.

Mikael Siversson
Curator of Palaeontology
Western Australian Museum

CRETACEOUS TIMELINE

The Cretaceous Period (145–66 million years ago) is just a fraction of the Earth's long history which stretches back over 4.5 billion years.

By the start of the Cretaceous most major groups of animals, plants and more simple life forms that we know today had evolved. During the Cretaceous, flowering plants first appeared, the ichthyosaur group died out and many individual species of plants and animals evolved and became extinct. The end of the Cretaceous is marked by a mass extinction event which many groups of organisms, most famously the non-avian dinosaurs, did not survive.

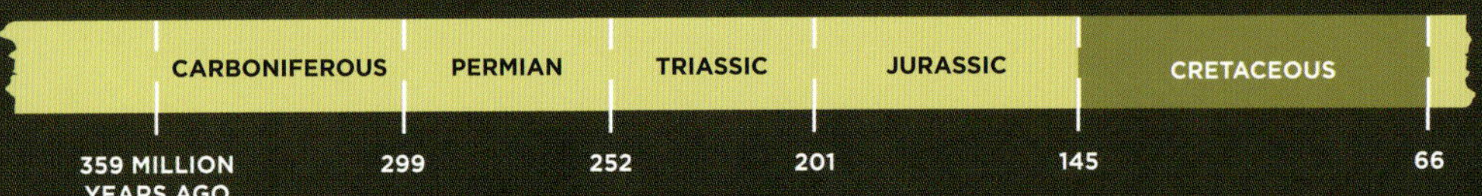

66 million years ago the impact of a massive asteroid or comet lead to the extinction of many groups of plants and animals including the non-avian dinosaurs.

NON-AVIAN DINOSAURS

AVIAN DINOSAURS

PTEROSAURS

CROCODYLIFORMS

PLESIOSAURS

SQUAMATA

ICHTHYOSAURS

TESTUDINES

SYNAPSIDS

AMPHIBIA

CHONDRICHTHYES

INSECTA

SPERMATOPHYTES

JURASSIC

Amargasaurus
Dwarfed sauropod from Argentina
131 MILLION YEARS AGO

Carcharodontosaurus
Huge predator from Morocco
108–98 MILLION YEARS AGO

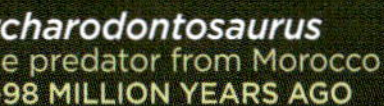
Spinosaurus
Giant theropod from Morocco
108–98 MILLION YEARS AGO

Tyrannosaurus
Huge predator from North America
68–66 MILLION YEARS AGO

Velociraptor
Small, feathered predator from Mongolia
75–71 MILLION YEARS AGO

Minmi
Small ankylosaur from Australia
130–104 MILLION YEARS AGO

Styracosaurus
Horned dinosaur from North America
75–74 MILLION YEARS AGO

Muttaburrasaurus
Beaked dinosaur from Australia
104–102 MILLION YEARS AGO

Therizinosaurus
Enormous clawed maniraptor from Mongolia
70 MILLION YEARS AGO

Australovenator
Swift predator from Australia
93 MILLION YEARS AGO

Leaellynasaura
Small ornithopod from Australia
113 MILLION YEARS AGO

Protoceratops
Small, horned dinosaur from Mongolia
75–71 MILLION YEARS AGO

EXTINCTION

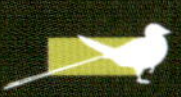
Confuciusornis
Crow-sized bird from China
125–120 MILLION YEARS AGO

BIRDS

Europejara
Early toothless pterosaur from Spain
126 MILLION YEARS AGO

EXTINCTION

Deinosuchus
Early giant relative of alligators
82–72 MILLION YEARS AGO

CROCODILES AND RELATIVES

Kronosaurus
Short-necked marine predator from eastern Australia
115–105 MILLION YEARS AGO

EXTINCTION

Lack of oxygen in deep oceans from extreme global warming killed off many prey species possibly causing extinction of the ichthyosaurs.

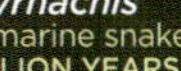
Pachyrhachis
Early marine snake from Israel
98 MILLION YEARS AGO

Snakes evolved around 105 million years ago and early forms still had visible back legs.

SNAKES, LIZARDS AND RELATIVES

MOSASAURS

Platecarpus
Marine lizard from North America
84–81 MILLION YEARS AGO

EXTINCTION

Platypterygius
Dolphin-like reptile from Australia
115–95 MILLION YEARS AGO

EXTINCTION

Archelon
Largest of all sea turtles from North America
74 MILLION YEARS AGO

TURTLES

Sinodelphys
Mammal closely related to marsupials from China
125 MILLION YEARS AGO

MAMMALS

Koolasuchus
Large predatory amphibian from Australia
120 MILLION YEARS AGO

Beelzebufo
Large frog from Madagascar
69 MILLION YEARS AGO

FROGS, SALAMANDERS AND RELATIVES

Cretoxyrhina
Giant shark that lived in many oceans of the time
105–80 MILLION YEARS AGO

SHARKS AND RELATIVES

Melittosphex
Early relative of bees and wasps from Myanmar
100 MILLION YEARS AGO

INSECTS

Araucaria
Conifer from Western Australia
145–125 MILLION YEARS AGO

Many early pine species evolved thick, fire-protective bark around 125 million years ago due to wide spread surface fires.

CONIFERS

ANGIOSPERMS

Leefructus
Early flowering plant from China
125 MILLION YEARS AGO

FLOWERING PLANTS

145 MILLION YEARS AGO | **EARLY CRETACEOUS** | **100 MILLION YEARS AGO** | **LATE CRETACEOUS** | **66 MILLION YEARS AGO** | **PALEOGENE**

DINOSAUR TYPES

Dinosaurs diversified into two main groups after they evolved some 230 million years ago in the Triassic Period.

One group, called Saurischian dinosaurs included the four-legged, long-necked, plant-eating sauropods and their relatives as well as the two-legged theropods. Most theropods were (and some continue to be) carnivores. The other group, called Ornithischian dinosaurs, were herbivores. The key differences between the two groups are in their hip bone and lower jaw structures.

Did you know?
Birds and some other closely related theropods, are classified as 'lizard-hipped' rather than 'bird-hipped'.

Saurischian dinosaurs

This group of dinosaurs have a hip structure that resembles that of lizards. The main portion of the pubis bone points forwards or downwards, and is located at an angle to the ischium bone.

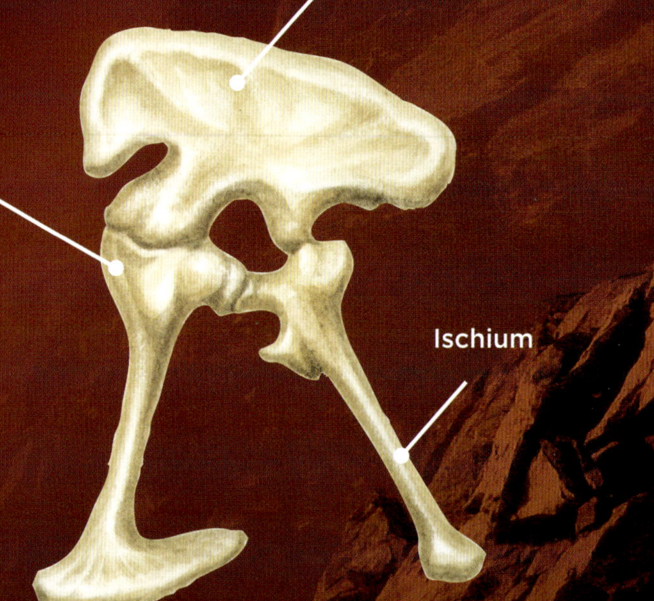

> Saurischian hip structure. (IMAGE: DORLING KINDERSLEY)

Ornithischian dinosaurs

This group of dinosaurs have a hip structure that superficially resembles that of birds. The main portion of the pubis bone points backwards and is located right along the ischium bone. They also have a bone in their lower jaw, called the predentary bone, that is toothless and forms a beak.

> **Ornithischian hip structure.** (IMAGE: DORLING KINDERSLEY)

< *Triceratops* skull indicating predentary bone.

(IMAGE: DAVID HERRAEZ CALZADA)

DRIFTING CONTINENTS

During the Cretaceous, as well as throughout the Earth's history, the continents have slowly moved position. This movement affects the evolution of life on Earth including dinosaurs.

How do they move?

The Earth's lithosphere (the crust and uppermost part of the hot mantle below) is divided into seven large and a few smaller, curved plates that move constantly, relative to each other. Continents have old, thick, but relatively light crust, whereas the ocean floor has young, thin and dense crust. Whenever the two types of crust collide, the denser oceanic crust will dive under the lighter continental crust and eventually melt deep in the interior of the planet. Some of the molten rock will rise as magma through the continental crust to form a chain of mountain-building volcanoes. New oceanic crust is continually formed at mid-oceanic ridges where two plates are moving away from each other.

Drifting dinosaurs

The slow reshaping of the Earth's crust had a major impact on the evolution of dinosaurs. In the Cretaceous, sea levels rose flooding the continents. By the Late Cretaceous dinosaurs became isolated on small islands in what is now central Europe and evolved into new and strange forms, like dwarfed sauropods.

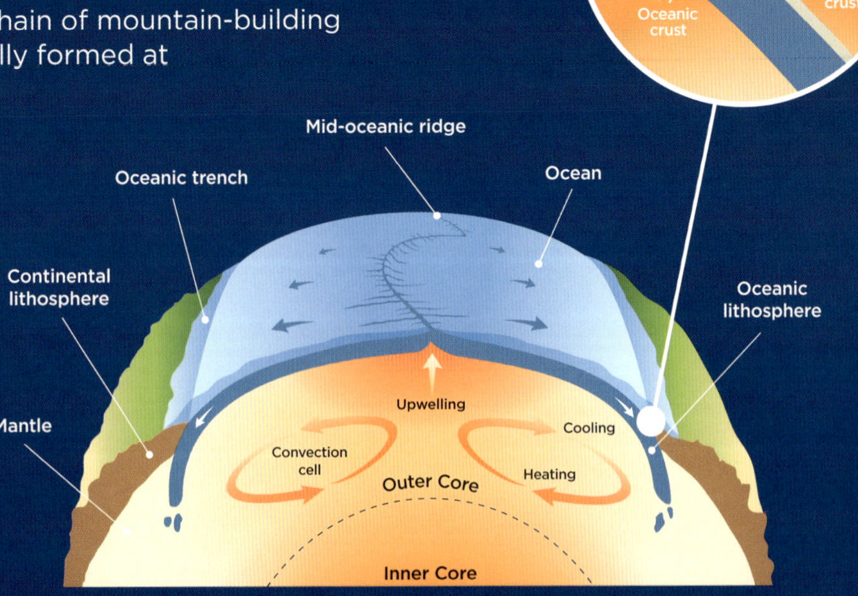

CRETACEOUS GREENHOUSE WORLD

In the Late Cretaceous, around 94 million years ago, the Earth experienced extreme global warming.

During this time the ocean floor changed greatly as the continents drifted apart. This increased plate movement and produced a surge in volcanic activity, both on the land and on the ocean floor. The increase in young and hot oceanic crust and a rise in water temperature resulted in a major sea level rise. A massive release of carbon dioxide (CO_2) from volcanoes and sea level rise are intimately linked to the vast oil and gas fields that date to this time.

Volcanic eruptions change the climate

Carbon dioxide was released in vast quantities by volcanoes erupting during this time. The rising sea and warmer climate flushed nutrients into the oceans. On the ocean floor, volcanic eruptions pumped iron and other micro-nutrients into the water. This resulted in enormous blooms of floating microscopic plants (called phytoplankton), while on the land the plants grew faster and both these events added more oxygen into the atmosphere. The high levels of oxygen in the atmosphere led to frequent and very hot wildfires across the planet.

Generating oil and gas

The greenhouse conditions ended as volcanic activity slowed and blooms of phytoplankton reduced levels of carbon dioxide in the atmosphere. As these blooms died, the oceans were depleted of oxygen. The build up of organic material on ocean floors during this period is the source of much of the world's oil and gas that today we are so dependent on.

> **Volcanic eruption.** (IMAGE: RAINER ALBIEZ)

AUSTRALIA IN THE CRETACEOUS

Australia in the Cretaceous was very different from today with volcanoes, shallow inland seas and great rift valleys.

The southern supercontinent Gondwana included Australia, South America, Africa, India, Antarctica and others. It had started to break up in the Jurassic some 160 million years ago but in the Early Cretaceous this intensified.

Early Cretaceous

A rapidly expanding seaway opened up between the western edge of Australia and India 130 million years ago. Also a deep opening, called a rift valley, was slowly forming between what is now the southern coast of Australia and Antarctica as these continents began to break apart. In the later part of the Early Cretaceous, up to 50 percent of the Australian continent was repeatedly flooded with shallow seas caused by rising global sea levels and vertical movements of its crust.

^ **The break up of Eastern Gondwana**
The globe above shows Australia in the Early Cretaceous, breaking away from Antarctica having separated from India. Much of the continent is covered by shallow seas.

Late Cretaceous

Around 95 million years ago, volcanic activity along the north-eastern edge lifted the continent in the east and the huge inland sea dried up. The sea was replaced by vast floodplains that provided an ideal habitat for sauropods and other dinosaurs. Many fossilised dinosaur remains are currently being uncovered in central Queensland particularly from the Winton Formation. The western edge of Australia remained underwater throughout much of the Late Cretaceous and marine rocks of this age are exposed in the Perth and Carnarvon Basins.

POLAR DINOSAURS

South-eastern Australia was once located near the South Pole and this impacted on the dinosaurs living there.

During the Cretaceous, the land that is now the southern Victorian coast was quite close to the South Pole. While the climate back then was warmer than it is today, this area still would have been freezing cold during long, winter nights. The mix of dinosaurs living in this area was unusual, with most species being small plant-eaters. Given their small size it is unlikely that they undertook annual migrations to escape the polar winter. Evidence indicates they dug burrows for shelter, which may have given them a competitive advantage over other dinosaurs.

Did you know?
The natural light display, that was visible in southern Victorian night skies at this time, is called the *aurora australis*.

Polar rift

Around 125–105 million years ago, Australia continued to slowly separate away from Antarctica. By this time the deep opening, called a rift valley, that started to form along the southern coast of Australia early in the Cretaceous included southern Victoria where these polar dinosaurs were living.

> **Downy dinosaurs**
To the right is an artist's impression of what Cretaceous polar dinosaurs may have looked like. While no direct evidence of downy skin has been found from Victorian dinosaur fossils, a similar small ornithopod known from China had skin with hair-like filaments that insulated it from extremes of temperature. It is unlikely that all the trees in the polar region lost their leaves in winter as this image shows. Most fossil trees from this time were evergreens.
(IMAGE: MASATO HATTORI)

LEAELLYNASAURA

Small, mid-Cretaceous ornithopods, like *Leaellynasaura*, have turned out to be quite remarkable dinosaurs.

Small ornithopods receive little mention in popular science literature compared to their famous meat-eating relatives. Despite a lack of deadly teeth and claws, they were remarkable — they could survive and even thrive in cold environments. Recent preliminary research suggests that *Leaellynasaura* might have had a super-long, flexible tail (longer than we have shown in our conservative reconstruction). This dinosaur lived in a polar region, and it is likely it had skin with some sort of hair-like filaments. By wrapping a long and fluffy tail around its body, *Leaellynasaura* could have endured cold snaps during winter.

SOUTHERN AUSTRALIA
(110 million years ago)

< ***Leaellynasaura amicagraphica.***
(IMAGE: PETER SCHOUTEN)

Fast growers

The leg bones of *Leaellynasaura* and other small ornithopods, found in the Early Cretaceous rocks of Victoria, come in all different sizes. From the bone structure, scientists have concluded that these dinosaurs grew very quickly during the first three years of their lives. After that, overall growth slowed down and varied according to seasons.

Big eyes

The skull of *Leaellynasaura* has large eye sockets, as do those of other small ornithopods found elsewhere, indicating they were active in low-light conditions. This dinosaur lived in the southern polar region and during winter the Sun would not have risen above the horizon for weeks on end. However the area would have been surprisingly well lit during the winter, even at night. This is because back then, as well as now, skies over polar areas have natural light displays, called aurora, and the snow reflects these as well as the light of the moon.

V **Polar night sky**
The brightness of the snow reflecting the light from this aurora below shows how it can be quite well lit at night in the polar zone.
(IMAGE: STRAHIL DIMITROV)

Leaellynasaura amicagraphica
NAMED BY RICH & RICH, 1989.

Diet:	Plants
Height:	0.4m
Length:	1.0m
Width:	0.2m

AUSTRALOVENATOR

This predator with large, deadly claws is known from one of Australia's best dinosaur fossil sites.

Australovenator was 5–6 metres long and a swift predator. It is based on the most complete skeleton of any Cretaceous meat-eating dinosaur from Australia. Its bones were found with fossilised sauropod, crocodile, turtle and fish bones as well as shells. Scientists think these animals were buried in clay at the bottom of an ancient billabong, in what is now central Queensland. At this fossil site, in the Winton Formation, dinosaur bones are being found every year.

NORTHERN AUSTRALIA
(93 million years ago)

^ *Australovenator wintonensis.*
(IMAGE: PETER SCHOUTEN)

Did you know?
Australovenator's teeth were not as big as those of tyrannosaurs of the same body size, but its arms were much larger and armed with deadly claws.

Family mystery

Australovenator belongs to the group of large-clawed carnivorous dinosaurs called megaraptors. This group has proved hard to place on the dinosaur family tree. Some scientists think they are closely related to spinosaurs, others say they are closer to allosaurs and yet others believe their closest relatives are tyrannosaurs and maniraptors. Whoever their relatives, fossils found in Queensland and Victoria, indicate megaraptors were the dominant group of meat-eating dinosaurs in Australia during the mid-Cretaceous.

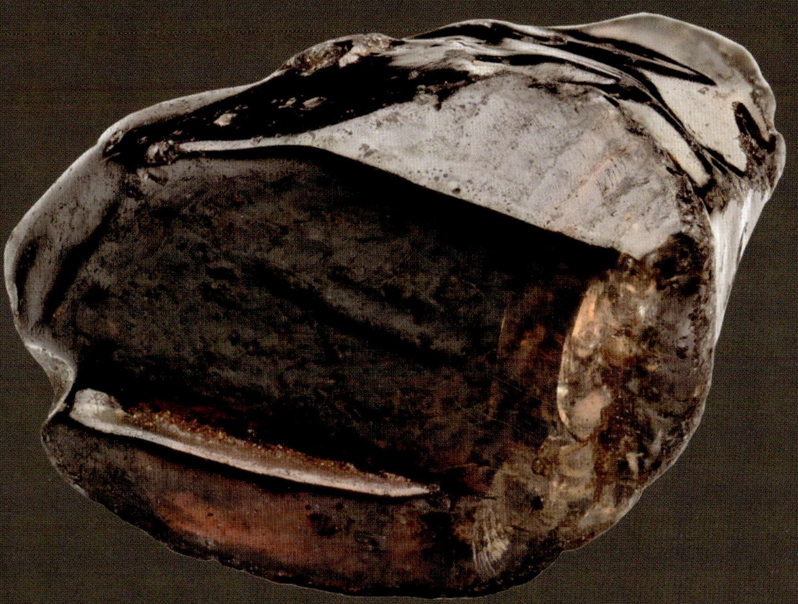

Change of age

Scientists have shown that *Australovenator* was almost 10 million years younger than previously thought. They did this by measuring the age of zircon grains, within the rocks of the Winton Formation that contained this fossil dinosaur. It is now believed that *Australovenator* was about 93 million years old.

< Brown Zircon crystal.
(IMAGE: INGEMAR MAGNUSSON)

Australovenator wintonensis
NAMED BY HOCKNULL et al., 2009.

Diet:	Animals
Height:	1.6m
Length:	5.0m
Width:	0.7m

MUTTABURRASAURUS

Although all Cretaceous dinosaurs (except for some emerging birds) lived on land, skeletons of some species have only been found in marine rocks.

This dinosaur is known from the discovery of a fairly complete skeleton, which is unlike most other Cretaceous dinosaur finds from Australia. The skeleton was found in 1963 near the town of Muttaburra in central Queensland. *Muttaburrasaurus* was quite large, measuring about 7 metres in length. It walked on all fours when feeding on understorey vegetation but ran on two legs. The bony bump on its snout may be related to how it made calls or its sense of smell.

NORTHERN AUSTRALIA
(105 million years ago)

< *Muttaburrasaurus langdoni.*
(IMAGE: PETER SCHOUTEN)

Did you know?
Several ornithischian dinosaur groups evolved large beaks and as they no longer needed front teeth, they were lost.

Marine finds

The original skeleton was found in rocks that formed in a marine environment about 105–103 million years ago. Two additional skulls, which may belong to a different species of *Muttaburrasaurus*, have also been found in slightly older marine rocks. This does not mean that *Muttaburrasaurus* was an aquatic animal, but reflects the fact that shallow seas covered much of eastern Australia at this time. Occasionally, a bloated carcass of a dinosaur would wash out to sea and float around for days.

SCAN ME

Beaked mouth

Though this dinosaur had no front teeth, this part of its upper and lower jaws were sheathed with keratin that formed a beak. *Muttaburrasaurus* did have back teeth and they were tightly packed, forming two continuous shearing blades on each side of the jaw. This allowed it to eat tough plant material.

> Skull of the North American iguanodontid *Saurolophus osborni* with the front end of the snout developed into a toothless beak.
(IMAGE: DIDIER DESCOUENS)

Muttaburrasaurus langdoni
NAMED BY BARTHOLOMAI & MOLNAR, 1981.

Diet:	Plants
Height:	2.3m
Length:	7.5m
Width:	1.2m

MINMI

The coastal plains of what is now Queensland were important habitats for these armoured dinosaurs.

NORTHERN AUSTRALIA
(105 million years ago)

Minmi is the best known of all armoured dinosaurs (ankylosaurs) from Gondwana thanks to a nearly complete skeleton found in north-central Queensland that is 105 million years old. *Minmi* was moderately sized, similar to ankylosaurs of the Early Cretaceous found on other continents. But it has bony rods attached to the back bone, called paravertebrae, which were unique and help scientists to separate it from other ankylosaurs. *Minmi* had some protection against theropods through the plates of bone in its skin.

Ankylosaurs may have been one of the more common types of dinosaur in Australia during the Cretaceous. Apart from *Minmi*, isolated ankylosaur teeth have been found in Victoria from the Early Cretaceous and in Queensland from the Late Cretaceous. Footprints found in Western Australia, dated to the Early Cretaceous and previously thought to belong to the closely-related stegosaurs, may actually have been made by ankylosaurs.

> *Minmi paravertebra.*
(IMAGE: PETER SCHOUTEN)

Did you know?
By the Late Cretaceous, ankylosaurs from North America and Asia were much larger and better armed than *Minmi*. They had developed a large bony club at the end of their tail, probably to protect them from attack by tyrannosaurs.

Marine finds

Like *Muttaburrasaurus*, all specimens of *Minmi* have been found in marine rocks. *Minmi* lived during a time when much of eastern Australia was covered by a shallow sea and, occasionally, carcasses would drift out to sea after floods. One skeleton of *Minmi* was found together with teeth of small bramble sharks. The sharks may have fed on the dead dinosaur as it laid upside-down on the sea floor.

Mummified *Minmi*
Some features of this *Minmi* skeleton below suggest that the dead dinosaur dried out completely and became mummified before it was washed out to sea. This specimen is also the best preserved armoured dinosaur from the Southern Hemisphere.
(IMAGE: QUEENSLAND MUSEUM)

Minmi paravertebra
NAMED BY MOLNAR, 1980.

Diet:	Plants
Height:	0.7m
Length:	2.5m
Width:	0.6m

LATE BLOOMERS

Flowering plants, which are so familiar to us today, evolved in the Cretaceous, blooming relatively late in the history of life.

The first fossil evidence of flowering plants, called angiosperms, is from the early Cretaceous about 130 million years ago. By the end of the Cretaceous flowering plants had diversified in an explosion of varieties. They ended the reign of the ancient conifers, cycads and ferns in dominating the landscape. This dramatic change in vegetation was one of the most significant moments in the history of life.

Special relationship

The interaction between flowering plants and the insects that pollinated them appears to have facilitated rapid diversification in both groups. This process is known as coevolution and may account for the amazing variety of flowers and insects we see today.

Magnificent magnolia

The ancestors of magnolias are among the most ancient of flowering plants, with a fossil record from about 100 million years ago. They appeared before the evolution of bees so perhaps their tough flowers evolved to encourage pollination by beetles.

> Modern 'Southern Magnolia'. (IMAGE: CLIFF COLLINGS)

Did dinosaurs eat grass?

Grasses are a specialised kind of flowering plant that are conspicuous and important in the world today. However they did not evolve until near the end of the Cretaceous, and didn't begin to dominate until much later. Some fossilised dinosaur dung contains what appear to be grass structures. So perhaps some of the very last dinosaurs may have had the opportunity to dine on grass.

V **Modern wild grass field.** (IMAGE: PITSANU KRAICHANA)

CRETACEOUS INSECTS

The Cretaceous period was a significant time for the evolution of insects.

Although many familiar insects such as beetles, flies and cockroaches already existed, the Cretaceous saw the appearance of several new groups that play important roles in the ecology of the world today.

The consumers

Step outside today and one of the first insects you are likely to encounter are ants. Ants play an important role in ecosystems as key consumers of other organisms. Despite their vast number and variety today, ants only appeared in the mid-Cretaceous about 90 million years ago, and remained quite rare until about 50 million years ago. There is some evidence suggesting that ancient ants had social nesting behaviours, just like modern ants.

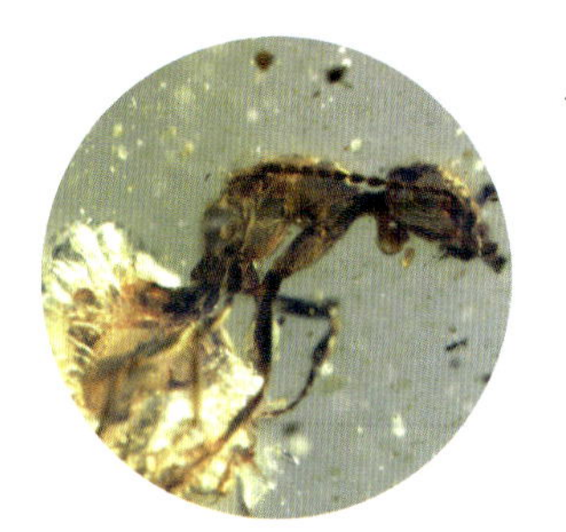

Cretaceous ant
Some of the best preserved fossil ants are found in amber, like this one from Myanmar. Amber is the fossilised resin of ancient trees.
(IMAGE: PHILLIP BARDEN / WWW.ANTWEB.ORG)

The pollinators

Bees are the most important pollinators of flowering plants in the modern world. Fossil bees are known from the mid-Cretaceous, not too long after flowering plants first appear in the fossil record. Another important group of pollinating insects, the nectar sucking butterflies, also evolved in the Cretaceous.

Modern bee
Bees transfer pollen as they fly between flowers. This is important for the reproduction of flowering plants.
(IMAGE: TSEKHMISTER)

DINOSAUR
DISCOVERY
LOST CREATURES OF THE CRETACEOUS

BIRDS ARE DINOSAURS TOO

When evidence mounted that some dinosaurs had feathers and other features once thought to be unique to birds, birds were reclassified as dinosaurs.

In the 1990s spectacular fossils of feathered dinosaurs were found in China. Fossils of extinct dinosaurs belonging to the group called maniraptors have features that are found in modern birds such as long forearms with modified wrist bones, wishbones and breastbones. Birds are now classified as maniraptoran dinosaurs. Some modern dinosaurs such as ostriches, have sharp claws on their fingers just like their ancestors.

> **Feathered Tyrannosaurs**
Yutyrannus from China was 9 metres long, a close relative of *Tyrannosaurus* and had simple, filament-like feathers that would have looked much like fur from a distance.
(IMAGE: SERGEY KRASOVSKIY)

^ **Egg brooding**
This skeleton of the emu-sized dinosaur *Citipati* is in a brooding position with the wings spread over the eggs in an attempt to protect them.
(IMAGE: AMERICAN MUSEUM OF NATURAL HISTORY)

Did you know?
Citipati and other oviraptor dinosaurs were almost more bird-like in appearance than were some of the true birds that lived during the Cretaceous.

EGGS, NESTS AND EMBRYOS

Extinct non-avian dinosaurs laid eggs like modern birds do.

Large dinosaurs like sauropods (long-necked dinosaurs), hadrosaurs (duck-billed dinosaurs) and therizinosaurs laid spherical eggs, whereas smaller theropods laid elongated ones.

Sauropod nests

Female sauropods dug an elongated pit and laid up to 28 eggs in it, most likely in one sitting. Most sauropod species had eggs with quite porous shells, so oxygen and carbon dioxide could pass through the shell. Their eggs were probably covered with sand or plant material to stop water loss through the shell.

> **Hatchling predator**
Several sauropod nests have been discovered in India with the skeletons of the ancient snake *Sanajeh.* Fossils such as this, of a *Sanajeh* curled around crushed eggs, indicate it preyed on baby dinosaurs. (IMAGE: JEFFREY A. WILSON et al.)

Oviraptor nests

A fossilised pelvis of a female oviraptor dinosaur found with eggs inside, shows that they produced two large, elongated eggs at a time. Their nests are bowl-shaped and contain up to 24 eggs arranged in a ring. These were probably laid over 2–4 weeks. The eggs were not very porous so they were unlikely to be buried. Instead the female, or perhaps the male, would sit on the nest.

Nesting colony of therizinosaurs
Therizinosaurs, the largest of the maniraptors, nested in colonies. Parental care of the hatchlings may have varied considerably from one group of dinosaurs to another and is difficult to determine from fossils.
(IMAGE: MASATO HATTORI)

THERIZINOSAURUS

This unusual dinosaur's enormous claws, shaped like giant scythes, could stop a tyrannosaur dead in its tracks.

With broad hips and a pot belly, this dinosaur was perhaps the strangest-looking of all dinosaurs. An adult *Therizinosaurus* is thought to have weighed about 6 tonnes, that's as heavy as a large *Tyrannosaurus rex*. It lived in the same area of today's Mongolia as the notorious *Velociraptor*, but around 5 million years later. By then a wetter climate had replaced the deserts in which *Velociraptor* lived. This later, fertile flood plain supported a much larger variety of dinosaurs, including giant forms which were absent in the deserts.

Did you know?
By examining the shape of the wrist and hip bones scientists have worked out that therizinosaurs are closely related to *Velociraptor*.

EASTERN ASIA
(70 million years ago)

< *Therizinosaurus cheloniformis.*
(IMAGE: PETER SCHOUTEN)

Strange dinosaur

Therizinosaurs appeared 120 million years ago and their remains have been found in Asia and North America. Because these dinosaurs had strange features it took scientists a long time to realise where they belonged in the dinosaur family. They belong to the theropod dinosaur group called maniraptors which also includes birds. All members of this group have long arms and therizinosaurs had hands equipped with the largest claws in the animal kingdom. Their small heads and the shape of their teeth show that therizinosaurs abandoned the meat-eating habits of their theropod ancestors and ate plants.

Handy in a fight

The main threat to *Therizinosaurus* was a huge tyrannosaur called *Tarbosaurus bataar*. Unable to outrun such a predator, *Therizinosaurus* would have stood its ground, towering over the tyrannosaur with metre-long claws ready to strike. A frontal attack on an adult *Therizinosaurus* may not have ended well for this tyrannosaur as it did not have the type of vision for judging distance very precisely that was necessary to avoid being stabbed.

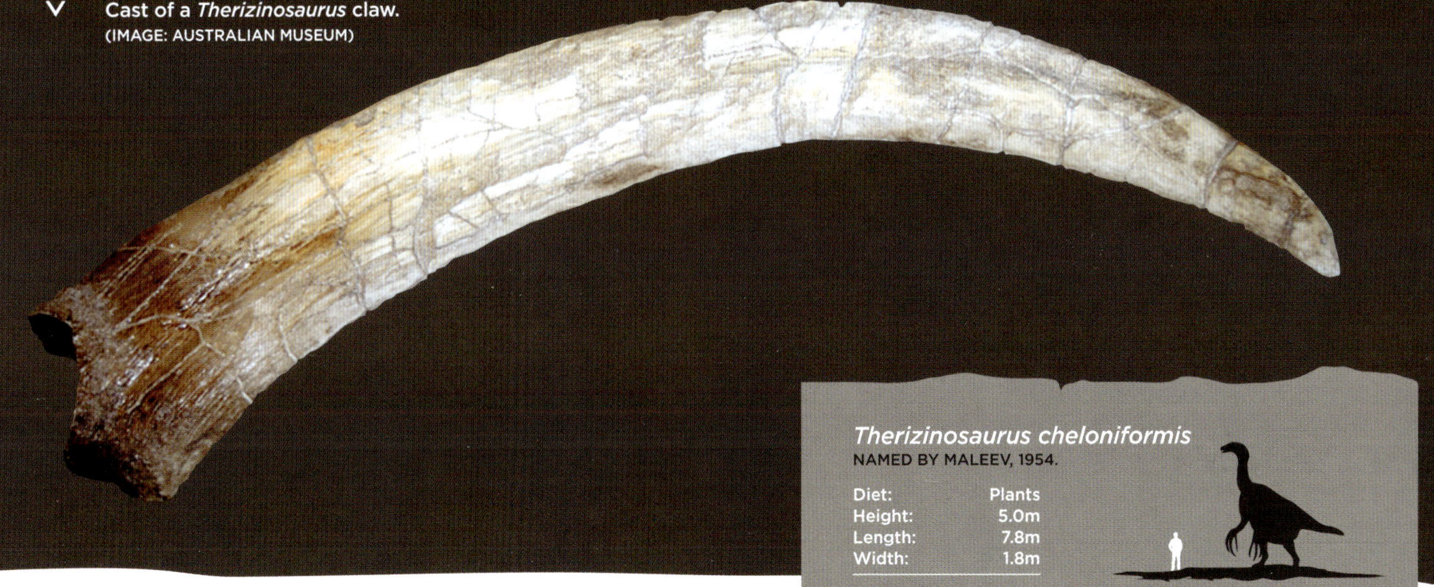

Cast of a *Therizinosaurus* claw.
(IMAGE: AUSTRALIAN MUSEUM)

Therizinosaurus cheloniformis
NAMED BY MALEEV, 1954.

Diet:	Plants
Height:	5.0m
Length:	7.8m
Width:	1.8m

Feathered predators

In 2007, evidence indicating *Velociraptor* had feathers was found on the forearm bones of a 75-million-year-old specimen. *Velociraptor* probably also preyed on small mammals and lizards, pinning them to the ground with its enlarged, sickle-shaped claw of the second toe. Some scientists believe *Velociraptor* and close relatives, also used their enlarged claws to climb trees.

> **Claw of *Velociraptor***
Cast from a real specimen from Mongolia that was around 75 million years old.

Beaked prey

Protoceratops, like all ornithischian dinosaurs, had a predentary bone that formed the beak of their lower jaw. The jaws of *Protoceratops* were lined with numerous, small teeth that formed a scissor-like tool that sliced up tough plants. Skeletons of *Protoceratops* have been found in large numbers suggesting they lived in herds and sometimes died together, perhaps during a flood.

> **Protoceratops nest**
The nest opposite with 15 hatchlings was probably buried by a migrating sand dune. It was found in 75 million-year-old rocks in the Gobi Desert, Mongolia. It indicates *Protoceratops* may have cared for their young.
(IMAGE: MARK WEICH)

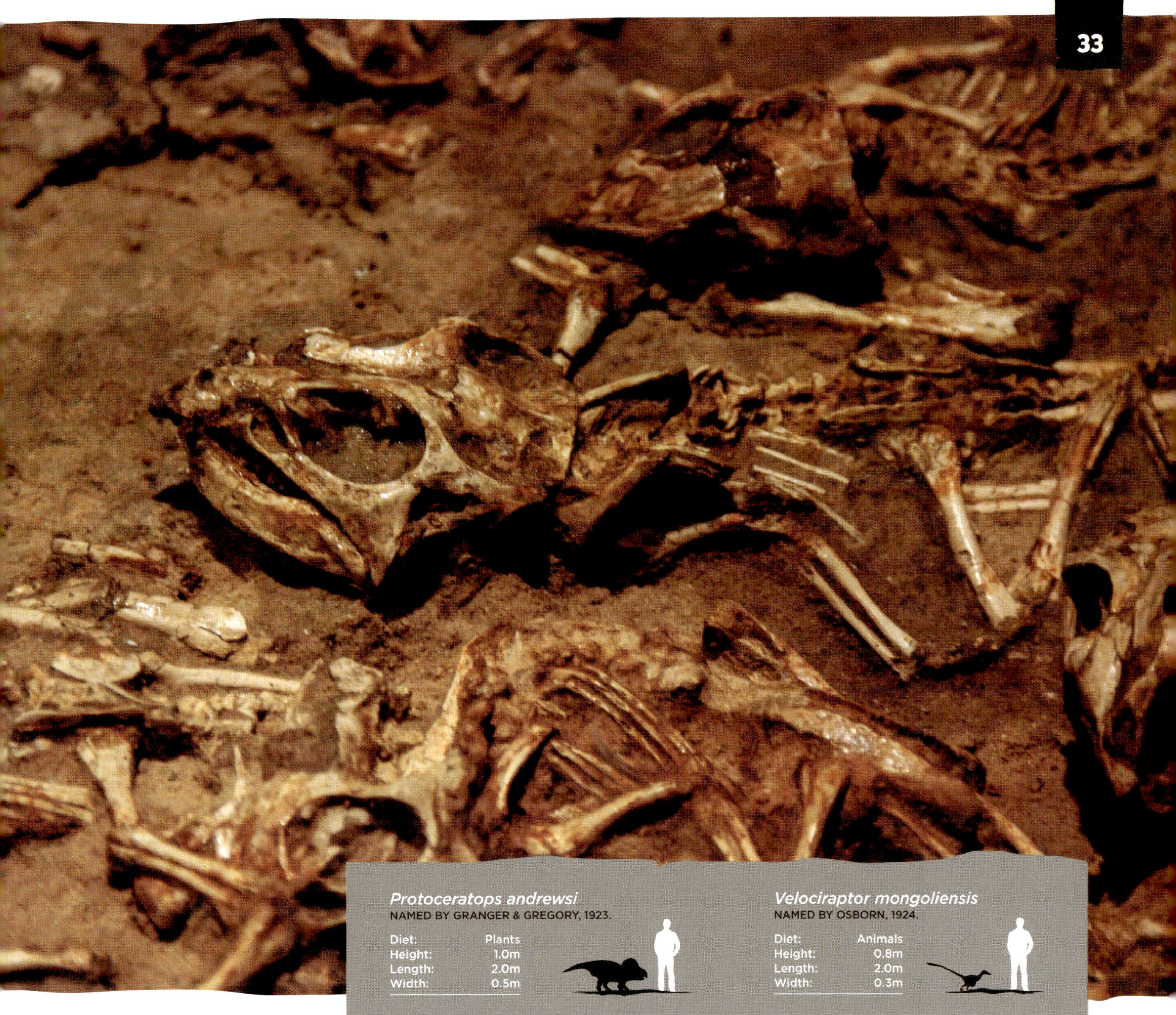

Protoceratops andrewsi
NAMED BY GRANGER & GREGORY, 1923.

Diet:	Plants
Height:	1.0m
Length:	2.0m
Width:	0.5m

Velociraptor mongoliensis
NAMED BY OSBORN, 1924.

Diet:	Animals
Height:	0.8m
Length:	2.0m
Width:	0.3m

DINOSAUR LOOK-A-LIKES

There is often confusion about what a dinosaur really is.

The frequent mention in newspapers of plesiosaurs as 'aquatic dinosaurs' and pterosaurs as 'flying dinosaurs' highlights this confusion. Although they are both prehistoric reptiles they are only distantly related to dinosaurs. Some even say crocodiles are dinosaurs but these living reptiles are only distant relations as well.

Marine reptiles
This plesiosaur, called *Zarafasaura oceanis*, lived in the oceans of Morocco in the Late Cretaceous. Some in this group of marine reptiles had shorter necks and larger heads.
(IMAGE: NOBUMICHI TAMURA)

Flying reptiles

Pterosaurs were spectacular and diverse animals. This one above belonged to the azhdarchid group which included the largest of all the pterosaurs. The biggest species in this group were as tall as a giraffe when standing on the ground and had a wingspan of up to 10 metres.

^ Airborne pterosaur of the Azhdarchidae family.
(IMAGE: MARK P. WITTON/SCIENCE PHOTO LIBRARY)

Family relations

Pterosaurs are more closely related to dinosaurs than are plesiosaurs, but they are still not dinosaurs. The hip socket, for example, is formed by bone in pterosaurs but marked by an open hole in dinosaurs. Pterosaurs, crocodiles and dinosaurs (including birds) belong to a group called archosaurs and share several unique features in their skeleton. They all have teeth set in sockets although both pterosaurs and dinosaurs eventually evolved toothless forms.

DINOSAUR FOOTPRINTS

Preserved dinosaur footprints capture unique moments in prehistoric time and reveal behaviour.

Footprints of extinct dinosaurs have been found on every continent, except Antarctica. These reveal behaviour that may be difficult to deduce, or prove, from studying fossilised bones alone. Most of what we know about Cretaceous dinosaurs from Western Australia is based on trackways preserved in the Broome Sandstone. Unlike fossil bone, footprints are not easily destroyed by acidic groundwater and may be preserved in weathered rocks where the bones of dinosaurs are long gone.

Examining footprints

By examining individual footprints and trackways, scientists have found that dinosaurs that walked on four legs walked much like quadrupedal mammals, with their limbs directly underneath the body. Trackways rarely show any evidence of tail drag marks which is why dinosaurs are now thought to have held their tails well off the ground. By measuring the stride length and the size of dinosaur footprints, scientists can estimate the height of the dinosaur at the hips and how fast it was walking or running.

> Cast of a 100 million-year-old bipedal footprint.
(IMAGE: SOUTHERN METHODIST UNIVERSITY)

> **Two-legged tracks**
Below are bipedal dinosaur tracks from the mid-Cretaceous discovered at Dinosaur Ridge, near Morrison, Colorado, USA.
(IMAGE: JIM HARRIS)

FROM DINOSAUR TO FOSSIL

Bacteria are part of the fossilisation process, turning dinosaur bones into stone.

Cretaceous dinosaur fossils are mainly found in areas that were flood plains during this time. Dinosaurs often drowned and were buried under the sand and mud (sediment) that spread over the flat landscape during flooding. Once buried, bacteria feeding on the decaying bones changed the chemistry in the surrounding sediment. This caused minerals in the ground water to form crystals within the bones, slowly turning them to stone. Over time, the weight of sediments spread over the flood plain caused it to sink and new layers of sediment to build up over the fossilised bones.

Radioactive dinosaurs

In some sedimentary rocks, uranium in the ground water can form crystals in the spaces within dinosaur bones, making them radioactive. The Morrison Formation in the USA from the Late Jurassic is well known for its radioactive dinosaur bones.

> **Yellow bones**
> This Late Jurassic dinosaur bone has been mineralised with the yellowish, radioactive uranium mineral called carnotite.
> (IMAGE: JAMES ST JOHN)

DINOSAUR POO

Dinosaurs feeding habits can be revealed by studying fossil dung called coprolites.

Unfortunately, in most cases, coprolites cannot be linked to specific species. However in 1998 scientists described a coprolite weighing more than 7 kilograms from 68–66 million-year-old rocks in central Canada. The dung is packed with broken bone fragments from what appears to be a juvenile hadrosaur. The only theropod dinosaur, known from fossils in that particular area, which was large enough to have produced such a sizeable poo is *Tyrannosaurus rex*.

T. rex digestion

The mass of highly fragmented bones in this dung indicates that *T. rex* had an exceptionally strong bite, crushing the bones to pieces. It also reveals that the digestive system of *T. rex* was quite different to that of living crocodiles, also meat-eating reptiles, which have highly acidic digestive juices that dissolve the hard, mineralised parts of the bones.

> Bone fragments found in *Tyrannosaurus rex* coprolite.
(IMAGE: ROYAL SASKATCHEWAN MUSEUM)

STOMACH STONES

Fossilised dinosaur skeletons have been found with stones (gastroliths) in their stomach region.

Birds swallow sharp, rough stones that help grind up food in their muscular second stomach, called a gizzard. This ensures they extract as much nutrition from their food as possible. Crocodiles also swallow stones, as did extinct long-necked plesiosaurs, but the function of these stones is unclear. The stones may help with controlling buoyancy but the total weight of stones in these animals is very small relative to the whole weight of the animal.

Theropod stones

Oviraptor skeletons have been found with gastroliths that have a rough surface, similar to the gastroliths of modern birds. As well, these fossils have virtually identical amount of stones relative to body size as modern birds. This is evidence that *Oviraptor* had a gizzard and that this feature is not unique to birds but rather is a more basic theropod character (like feathers) that is retained in birds (modern theropods).

Sauropod stones

It is quite rare to find sauropod skeletons with gastroliths. Those that have been found have gastroliths with smooth, polished surfaces as well as a small amount of stones relative to their body size. So scientists think these giant dinosaurs lacked a gizzard-type stomach.

> ***Psittacosaurus* skeleton**
The gastroliths of this plant-eating dinosaur, from the Early Cretaceous of Mongolia, have been preserved. This dinosaur's teeth were not suitable for grinding and gastroliths would have helped with digesting its food.
(IMAGE: RYAN SOMMA)

SPINOSAURUS

At an estimated maximum length of roughly 15 metres, *Spinosaurus* was the largest of all known theropod dinosaurs.

This enormous, strange-looking theropod prowled the coastal plains of northern Africa, 108–98 million years ago. The shape of its teeth and jaws indicates that *Spinosaurus* included fish in its diet, and a fossil of a partial skull was found with a fish vertebra lodged in a tooth socket. The rivers and estuaries of northern Africa were teeming with the giant sawfish called *Onchopristis* and other fish. The sheer size of an adult *Spinosaurus* probably enabled it to prey on *Onchopristis* up to several metres in length, while at the same time giving it some immunity to attacks by *Carcharodontosaurus*.

NORTHERN AFRICA
(100 million years ago)

^ *Spinosaurus aegyptiacus.*
(IMAGE: PETER SCHOUTEN)

Croc-like mouth

Spinosaurus teeth are similar to teeth of crocodiles and their close relatives and have been misidentified as such in the past. The shape of their jaw is also similar to that of several large species of crocodiles and their close relatives. These conical teeth and crocodile-like jaw shape helped *Spinosaurus* to secure a better grip on slippery prey.

Sail back

The sail-like structure on this dinosaur's back is formed by extremely tall spines of the vertebrae. Some scientists suggest the sail was for display, while others believe *Spinosaurus* used its sail to regulate body heat. Another possibility is that the sail served as internal support for a hump of fat built up during times of abundant food supply.

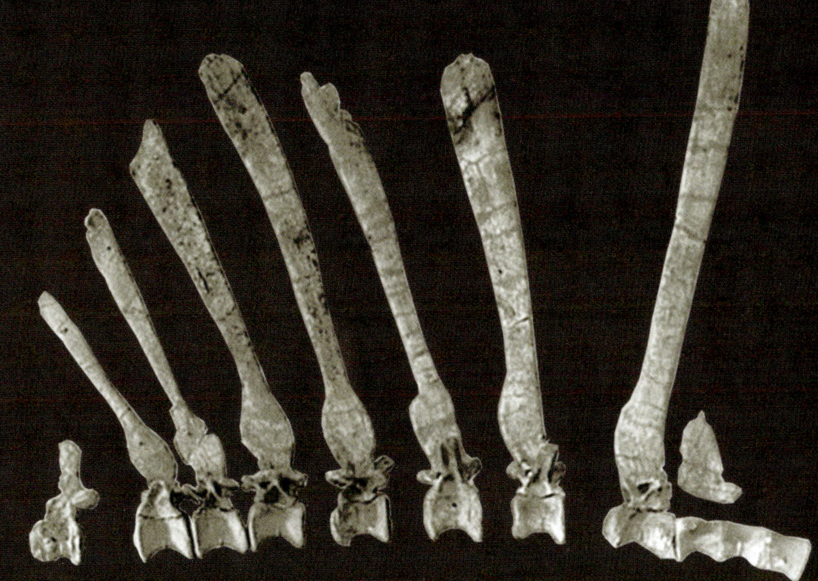

> *Spinosaurus* vertebrae with tall spines.
(IMAGE: WASHINGTON UNIVERSITY)

Spinosaurus aegyptiacus
NAMED BY STROMER, 1915.

Diet:	Animals
Height:	5.4m
Length:	15.0m
Width:	2.0m

CARCHARODONTOSAURUS

Packs of these large carnivores may have hunted huge sauropods.

Although cooperative hunting is difficult to prove in extinct animals, it is possible that *Carcharodontosaurus* hunted in packs. Their knife-like teeth with serrated edges could slice through the tough hide of very large dinosaurs. A bite from a *Carcharodontosaurus* would cause extensive bleeding and leave toxic bacteria in the wound. A pack may have been able to attack and by delivering slashing, toxic wounds, weaken and eventually bring down even the largest sauropods.

NORTHERN AFRICA
(100 million years ago)

^ *Carcharodontosaurus saharicus.*
(IMAGE: PETER SCHOUTEN)

Did you know?
Carcharodontosaurs from the mid-Cretaceous grew to a size similar to *Tyrannosaurus rex* that lived in the Late Cretaceous.

Widespread carnivores

This dinosaur and its close relatives belong to a group known as carcharodontosaurs. Most fossils of these dinosaurs have been found in 100–90 million-year-old rocks in Patagonia, Argentina, and Morocco. However fossils of *Acrocanthosaurus*, have been found in 115–110 million-year-old rocks from North America. This one had enlarged spines on its back vertebrae, but not nearly to the same length as in *Spinosaurus*.

Predators

This dinosaur lived at the same time and in the same place as *Spinosaurus*. It was almost as large as *Spinosaurus* and could have taken over their kills. Adult *Carcharodontosaurus* may have been a major predator of young and old *Spinosaurus*.

˅ *Carcharodontosaurus* encroaching on a *Spinosaurus*' kill.
(IMAGE: SERGEY KRASOVSKIY)

Carcharodontosaurus saharicus
NAMED BY STROMER, 1931.

Diet:	Animals
Height:	4.3m
Length:	11.0m
Width:	2.0m

BURROWING DINOSAURS

In 2007, scientists found the first evidence of a burrowing dinosaur.

They found the bones of an adult and two juveniles called *Oryctodromeus cubicularis*, a small ornithopod dinosaur. These bones were in the chamber of a filled burrow, preserved in 95-million-year-old rocks in Montana, USA. Several features of the adult skeleton are consistent with habitual burrowing.

Australian burrowing dinosaurs

Similar burrows attributed to digging activities by small ornithopod dinosaurs, such as *Leaellynasaura*, have been found in 110-million-year-old rocks in Victoria. This area was dominated by small ornithopod dinosaurs which is quite unusual. It has been suggested that some, if not most, of these made dens for shelter during the cold winter months.

Korean burrowing dinosaurs

Koreanosaurus boseongensis, described from 85 million-year-old fossils found in South Korea, was a small ornithopod with a skeleton highly adapted to digging. The arms and shoulder bones are extremely robust with huge attachment surfaces for muscles used in digging. The hind legs were rather short and sprawling, features also seen in modern mammals adapted to digging.

Did you know?
Oryctodromeus cubicularis **means 'digging runner of the lair'.**

> *Oryctodromeus* burrow.
(IMAGE: MARK HALLETT/NATIONAL GEOGRAPHIC CREATIVE)

BITE MARKS

Fossil bones of Cretaceous dinosaurs sometimes have bite marks inflicted by theropods which can tell us about how these predators fed.

Most of these bite marks are the result of kills or scavenging. In rare cases, the attacked dinosaur survived long enough for new bone growth to occur. Bite marks are not only found on the bones of herbivorous species but also on theropod bones. For example, skulls of tyrannosaurs are often found with bite marks inflicted by other tyrannosaurs.

Feeding habits

The pattern and distribution of bite marks reveal clues about the hunting techniques and feeding habits of predators. *Tyrannosaurus rex* hunted mainly hadrosaurs and *Triceratops* which was one of the largest and most dangerous of the horned dinosaurs. New research suggests that, after killing a *Triceratops*, *T. rex* often ripped its head off, perhaps to gain access to the extensive neck muscles.

> *Tyrannosaurus rex* attacking a *Triceratops*.
(IMAGE: GOW NIMMANHAEMIN)

Did you know?
Dinosaurs continuously replaced old teeth with new ones, like sharks and crocodiles today. Once a tooth was loose, it would easily fall out during feeding. Shed teeth from various theropods are commonly found near dinosaur skeletons with theropod bite marks.

HOW BIG COULD THEY GROW?

The largest theropod dinosaurs are estimated to have a maximum body size of around 12–13 metres.

Almost all fossil dinosaur skeletons are incomplete so their size is estimated by extrapolating from the bones that are found. *Torvosaurus* is arguably one of the largest Jurassic theropods. Its skull was estimated to be over 1.5 metres long based on a very large jaw bone found in 150-million-year-old rocks from Portugal. This skull is similar in size to the largest carcharodontosaur skull from the mid-Cretaceous and the largest tyrannosaur skull from the Late Cretaceous. A theropod skull of 1.5 metres corresponds to a body length of roughly 12–13 metres.

Limits to growth

When a body doubles in size, it becomes eight times heavier but its muscles and bones are only four times stronger. The only way to compensate for this effect is to increase the relative thickness of the bones and the muscles, which would make the theropod even heavier and place additional strain on the lungs and heart. So it seems that growing to about 12–13 metres is about the limit for these dinosaurs.

Did you know?
Spinosaurus grew a bit bigger than the 12–13 metres length that seems to be the general theropod size limit. It could afford to, as it probably spent much of its time simply waiting for fish to swim past rather than actively trying to hunt down other dinosaurs.

6m

12m

V ***Australovenator wintonensis*** **and** ***Tyrannosaurus rex.***
(IMAGE: PETER SCHOUTEN)

CRETACEOUS AMPHIBIANS

Giant frogs from Madagascar along with other large amphibians living in the Cretaceous may have eaten small dinosaurs.

Predatory frogs

Beelzebufo ampinga, nicknamed the 'devil frog', is known from fossils recently discovered in 72–66 million-year-old rocks from Madagascar. This frog is closely related to modern South American horned frogs (*Ceratophrys*) which are voracious and fearless ambush predators that feed on frogs, mice, lizards, snakes, birds and larger insects. *Beelzebufo* had a huge mouth relative to its body size and this indicates that it too ate mainly vertebrates. Just about anything small enough to fit into its gigantic mouth was potential prey, including baby dinosaurs.

Australian giant amphibian

A huge cool-water amphibian, called *Koolasuchus cleelandi*, lived in Australia some 120 million years ago. Fragments of its jaws and other bones found in Victoria indicate it could grow to 4–5 metres in length and weigh a tonne. These amphibians were top aquatic predators, similar to crocodiles today. *Koolasuchus* became extinct towards the end of the Early Cretaceous when temperatures were rising and this gave crocodiles the opportunity to take over.

Modern giant amphibians

The largest living amphibians, giant salamanders from China and Japan, grow to 1.5–1.8 metres long. Like their extinct and distant relative *Koolasuchus*, they thrive in fast-flowing waterways that are too cold for crocodiles. Cold water holds more oxygen than warm water, and these modern, scaleless amphibians absorb oxygen from the water through their skin.

Did you know?
Koolasuchus had a mouth so big it could swallow a *Leaellynasaura* dinosaur whole and it probably could have swallowed you too.

V *Beelzebufo ampinga* eating a small dinosaur.
(IMAGE: TODD MARSHALL)

TYRANNOSAURUS

Huge teeth and a bone-crushing bite enabled a fully-grown *Tyrannosaurus rex* to hunt and kill extremely dangerous prey.

NORTH AMERICA
(66 million years ago)

T. rex was the largest species in the tyrannosaur group and one of the last to evolve, near the end of the Cretaceous. While even larger meat-eaters belonging to other dinosaur groups have now been found, scientists think *T. rex* had, by far, the most powerful bite of any dinosaur. Its broad skull was made of very tough bones and had deep jaws with massive banana-shaped teeth to withstand such bone-crushing bite force. These features, as well as a strong neck and forward facing eyes, indicate *T. rex* was an active hunter of dangerous prey. Its highly-developed sense of smell suggests this dinosaur scavenged as well.

^ *Tyrannosaurus rex.*
(IMAGE: PETER SCHOUTEN)

Did you know?
While *T. rex* probably did most of its hunting alone, other tyrannosaurs may have hunted in packs attacking huge sauropods.

Female fossils

Scientists have been able to identify a *T. rex* fossil as female based on a feature of its living relatives. Female ostriches and emus develop a certain type of bone tissue, called medullary bone, inside their leg bones every time they produced a batch of eggs. A *T. rex* has been found with evidence of medullary bone tissue inside its leg bones. So the specimen can be identified as a female that died while she was producing eggs.

Massive meat-eaters

T. rex had a similar body size and shape to another super-predator, *Carcharodontosaurus* from the mid-Cretaceous, but they are not closely related. Distantly related animals that have a similar position in the food chain tend to evolve a similar body shape over time. Before carcharodontosaurs became extinct around 90 million years ago, tyrannosaurs lived as lesser predators and were relatively small.

> **The skull and banana-shaped teeth of the *T. rex* designed for crushing the bones of its prey.**
> (IMAGE: DMITRY IDANOV)

Tyrannosaurus rex
NAMED BY OSBORN, 1905.

Diet:	Animals
Height:	4.6m
Length:	12.0m
Width:	1.8m

AMARGASAURUS

Amargasaurus lived 130 million years ago in what is now Argentina, and represents a unique group of dwarfed, short-necked sauropods.

SOUTH AMERICA
(130 million years ago)

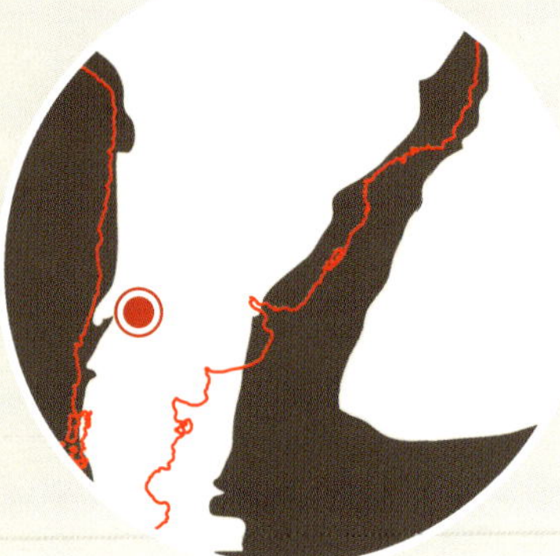

Amargasaurus belongs to a peculiar group of sauropods from Gondwana, known from the Late Jurassic to the Early Cretaceous. They are related to *Diplodocus*, a giant, long-necked sauropod from the Late Jurassic of North America. However they went down a different evolutionary path to their ancestors and evolved a much shorter neck and a drastically reduced overall body size.

< ***Amargasaurus cazaui.***
(IMAGE: PETER SCHOUTEN)

Did you know?
Sauropods seem to have had large air sacs inside their bodies, just like birds. These air sacs extended into holes in the back bones, making these bones both strong and light.

Long and short of it

Sauropods followed several evolutionary paths (often combined) to increase or decrease the length of their necks. Some evolved more vertebrae in the neck, others had the original number of vertebrae but each one became either longer or shorter, and a few had the most forward of their body vertebrae transform into neck vertebrae.

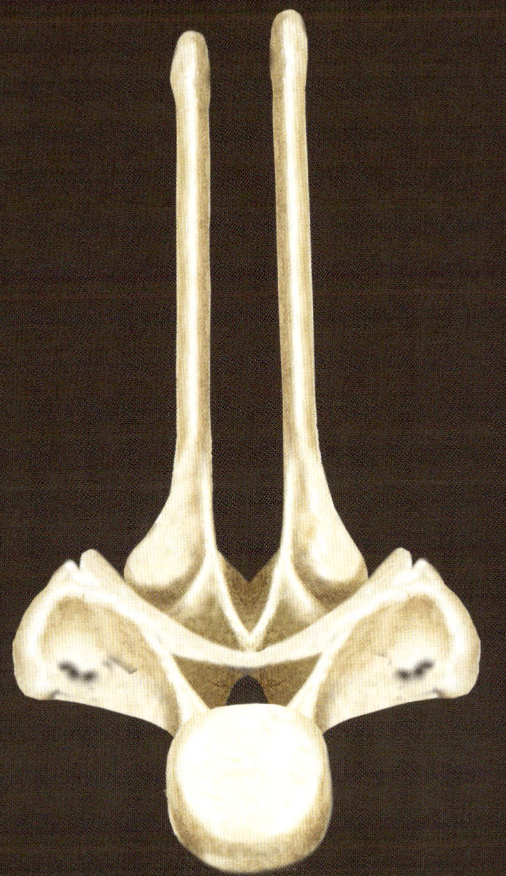

The size of the spines on the neck vertebrae and the way they are arranged, prevented these dinosaurs from raising their necks vertically. This restricted movement, as well as the short length of their neck, limited these dinosaurs to feeding on understory plants. It is possible that the small size of these sauropods was linked to their more limited food source.

> **Spine speculation**
> This illustration shows how the spine on each of the neck and back vertebrae of *Amargasaurus* is greatly elongated and divided into two. Some scientists think the spines supported a sail, however we have shown an alternate view, that each spine had an individual covering. It is possible that the spines were used for defence. By bending the neck, *Amargasaurus* would be able to direct the spines towards an approaching predator.

Amargasaurus cazaui
NAMED BY SALGADO & BONAPARTE, 1991.

Diet:	Plants
Height:	2.5m
Length:	10.0m
Width:	1.2m

END OF AN ERA

An asteroid or comet impact unleashed a series of cataclysmic events, ending the reign of dinosaurs.

Some 66 million years ago, an asteroid or comet, estimated to have been up to 10 kilometres across, smashed into what is now a coastal area of Mexico. The impact generated an enormous shock wave and tidal waves reached far inland. Glowing debris from the impact was catapulted back into space only to be pulled back by the Earth's gravity. The returning debris generated an intense heat pulse as it rained down which may have been strong enough to cause global fire storms.

Mass extinction

The initial mayhem was followed by darkness and falling temperatures as dust and condensed sulphuric acid spread throughout the atmosphere, drastically reducing sunlight for years. The photosynthesis of plants and algae slowed down and many organisms, including all dinosaurs except birds, died out.

Initial impact
This reconstruction shows the glowing debris being catapulted back into space just after the asteroid or comet struck.
(IMAGE: SOLAR SEVEN)

AGE OF THE MAMMALS

Mammals began to thrive after the end of the Cretaceous when all dinosaurs, except toothless birds, became extinct.

Mammals evolved before the Cretaceous and during this period most were very small. This is probably because they were hunted intensely once they reached a certain size, most likely by small theropod dinosaurs like *Velociraptor*. Mammals living in the Cretaceous belonged to the several extinct groups, monotreme (egg-laying) and marsupial (pouched) groups, and possibly the placental group. After the Cretaceous, mammals increased rapidly in size and took over territory vacated by dinosaurs.

Mammal history

Mammal-like reptiles had been dominant in the Permian Period (299–252 million years ago). Then during the Triassic, 230 million years ago, dinosaurs appeared and gradually replaced these mammal-like reptiles. However one group of mammal-like reptiles, called cynodonts, survived. They became smaller, probably to avoid attention from dinosaurs, and eventually gave rise to early mammals.

Rapid growth
This large bear-like but herbivorous mammal, called *Titanoides*, lived 60 million years ago, only a few million years after the extinction of non-avian dinosaurs.
(IMAGE: ROMAN UCHYTEL)

EARTH DURING THE CRETACEOUS

The world 66 million years ago.

Earth during the Cretaceous Period experienced major sea level rises that flooded land masses. The isolation of many dinosaur groups contributed to the high diversity of body shapes in Cretaceous dinosaurs.

IMPACT ZONE

145 MILLION YEARS AGO

90 MILLION YEARS AGO